Reaching for the Stars

Written and Illustrated
by Anna Brusoe

Morgantown, W.V. 2021

Reaching for the Stars

ISBN: 978-0-578-30407-6

Cover design by Anna Brusoe

Back cover photo by Russell Purdum

nna Brusoe
mail: reachingforthestars.projectnews@gmail.com

nis is a work of non-fiction. Every attempt has been ade to ascertain the accuracy of factual material in this ook.

Table of Contents

Name Pronunciation Guide

Katherine Johnson
(kah·thuh·rin john·sohn)

Jane Luu
(jehn luh·ooh)

Mary Golda Ross
(meh·ree gol·da ross)

Ellen Ochoa
(eh·lun oh·choh·ah)

Cecilia Payne-Gaposchkin
(suh·cee·lee·ah payn guh·paash·kin)

Key Definitions

Asteroid: (a·str·oyd) A space rock that orbits around the sun.

Astronaut: (a·struh·naat) A person who travels to space.

Astronomy: (uh·straa·nuh·mee) The study of space and everything in it.

Engineer: (en·juh·neer) A person who designs and creates various products.

Hydrogen: (hai·druh·jn) The most abundant element in the universe.

Key Definitions

Kuiper Belt: (kai·pr belt) A place located beyond the planets in our Solar System where icy rocks orbit.

NASA: The National Aeronautics and Space Administration is an agency that works with space flight.

Solar system: The Sun and everything that orbits it.

S.T.E.M.: S.T.E.M stands for Science, Technology, Engineering, and Math.

Katherine Johnson

1918-2020

I like to learn. That's an art and a science.

Katherine Johnson

Katherine Johnson was from the rural Appalachian town of White Sulphur Springs, West Virginia. She was born in 1918. She had an interest in math and space from a young age and didn't let other people stop her from learning. Excelling in school, she finished her studies early.

Katherine Johnson

When she was older, Katherine faced many barriers because of her skin color! Despite this, she became one of the few African-American women during that time to work at NASA. In addition, because of her curiosity, she started to attend meetings that women hadn't attended before.

Katherine Johnson

Using her calculations, NASA was able to successfully send people to space. Without her, NASA might not have been able to put a person on the moon when they did! We still use her mathematical calculations today for space exploration.

Katherine Johnson

Katherine's influence didn't just extend to space travel. She was a pioneer for women and African-Americans and has received numerous awards for her work and bravery.

Jane Luu

b. 1963

Many more discoveries are waiting to be made in our solar system.

Jane Luu

Jane Luu was born in Vietnam during the Vietnam War. After immigrating to the United States, she worked hard in school and received an award to study at a reputable university. Once she finished school, she continued learning about space, especially comets and asteroids.

Jane Luu

Along with her colleague, astronomer David Jewitt, Jane set out to see what might lie beyond the planets. Their hard work paid off in 1992, when they discovered the Kuiper Belt, a place where comets and asteroids are stored.

Mary Golda Ross

1908-2008

Math was more fun than anything else. It was always a game to me.

Mary Golda Ross

Mary Golda Ross was one of the first Native American engineers. She was born in 1908 in Oklahoma as a member of the Cherokee Tribe. She was fascinated by math, and became a math and science teacher.

Mary Golda Ross

During World War II, she received a job at a company that designed spaceflight vehicles. She loved it and did so well that the company permanently hired her as an engineer, even though at the time it was very unusual for a Native-American woman to be hired.

Ellen Ochoa

b. 1958

Don't be afraid to reach for the stars.

Ellen Ochoa

Ellen Ochoa was born in California. She loved math and science and dreamed about becoming an astronaut from a young age. As a Hispanic woman, she had to work extra hard to make her dream come true.

Ellen Ochoa

After overcoming gender and ethnicity barriers, Ellen became the first Hispanic woman in space. In addition, she went back to space three more times and spent almost a total of 1000 hours in space!

Ellen Ochoa

Following her last mission in 2002, she explored space from Earth by working on research and being a leader in the astronomy community. Ellen became the first Hispanic director of one of NASA's most important centers, the Johnson Space Flight Center.

Cecilia Payne Gaposchkin

1900-1979

Your reward for studying science will be the widening of the horizon as you climb.

Cecilia Payne-Gaposchkin

From England, Cecilia Payne-Gaposchkin had an interest in science from a young age. But in the early 1900s, women weren't encouraged to pursue science. That didn't stop Cecilia; she made many significant discoveries in science and astronomy.

Cecilia Payne-Gaposchkin

Her work lead to solving the puzzle that had stumped scientists of previous generations. She discovered that the Sun is made primarily of hydrogen. From this discovery, she concluded that most of the universe is composed of hydrogen.

Index

Index

ACKNOWLEDGEMENTS

I would like to thank everyone who has helped me with this book and project. Specia thanks to Shannan Davis, Mary Duran, Beth Nardella, Nghi-Anh Brusoe, Girl Scout Troop 51102, Girl Scout Service Unit 421, family, and friends.

This book was written in partial completion of the Girl Scout Gold Award, the highest award a Girl Scout can earn.

Citations

Definitions

"Astronaut." *Purdue University: College of Science: Astronaut,* n.d.www.purdue.edu/science/careers/what_can_i_do _with_a_major/Career%20Pages/astronaut.html. Accessed 29 June 2021.

"Hydrogen - Element Information, Properties and Uses: Periodic Table." *Royal Society of Chemistry*, n.d.www.rsc.org /periodic-table/element/1/hydrogen. Accessed 27 Jan. 2021.

"Kuiper Belt." *NASA*, 13 Nov. 2019, solarsystem.nasa.gov/solar-system/kuiper-belt/overview/. Accessed 27 Jan. 2021.

"NASA for Kids: Intro to Engineering." National Geographic Society, 9 Nov. 2012, www.nationalgeographic.org/ media/nasa-kids-intro-engineering/. Accessed 29 June 2021.

Redd, Nola. "What Is Astronomy? Definition & History." *Space,* 8 Sept. 2017, www.space.com/16014-astronomy.html. Accessed 27 Jan. 2021.

"What Exists Within the Reaches of Our Solar System?" National Geographic, n.d. www.nationalgeographic.com/ science/space/our-solarsystem/. Accessed 27 Jan. 2021.

"What Is an Asteroid?" NASA, 1 June 2020, spaceplace .nasa.gov/asteroid/en/.Accessed 27 Jan. 2021.

"What is NASA?" Ed. by Sandra May. NASA, 24 Sept. 2018, https://www.nasa.gov/audience/ forstudents/5-8/features /nasa-knows/what-is-nasa-58.html. Accessed 22 June, 2021.

Citations

Katherine Johnson

"Inspirational Quotes: Katherine Johnson | NASA Mathematician." *We are Tech Women,* 3 Oct. 2020, https://wearetechwomen.com/inspirational-quotes-katherine johnson-nasa-mathematician/. Accessed 29 June 2021.

Shetterly, Margot Lee. "Katherine Johnson Biography." *NASA,* 24 Feb. 2020, https://www.nasa.gov/content/katherine-johnson-biography. Accessed 3 March 2021.

Wild, Flint. "Who Was Katherine Johnson?" NASA, 30 Dec. 2016, www.nasa.gov/audience/forstudents/k-4/stories/nasa-knows/who-was-katherine-johnson-k4/. Accessed 27 Jan. 2021.

Jane Luu

"Dr. Jane Luu Public Lecture: The New Solar System on May 4, 2013." Youtube, uploaded by Toan Vo, 7 May 2013, https://www.youtube.com/watch?v=wFxs_hrkP5E.

"Jane X. Luu." *Kavli Prize*, 18 Apr. 2018, kavliprize.org/prizes-and-laureates/laureates/jane-x-luu. Accessed 27 Jan 2021.

Luu, Jane. Email to Anna Brusoe. 21 September 202.

Venkatraman, Vijaysree. "No Starry Eyed Astronomer." *Science Magazin*e, 19 October 2012,https://www.sciencemag.org /careers/2012/10/ no-starry-eyed-astronomer. Accessed 27 Jan 2021.

Mary G. Ross

Briggs, Kara. "Cherokee Rocket Scientist Leaves Heavenly Gift." *Cherokee Pheonix,* 18 Dec. 2008, https://www.cherokeephoenix .org/Article/Index/2470. Accessed 27 Jan 2021.

"Mary G. Ross." National Park Service, 21 Nov. 2019, https://www.nps.gov/people/mary-g-ross.htm. Accessed 27 Jan. 2021.

Citations

Mary G. Ross(Continued)

"Who was Mary G. Ross? Facts and Quotes About the First Female Native American Engineer Celebrated in a Google Doodle." *Newsweek,* 9 Aug. 2021, https://www.newsweek.com/mary-g-ross-google-doodle-1064895. Accessed 29 June 2021.

Ellen Ochoa

"Ellen Ochoa." *National Science Board,* n.d. https://www.nsf.gov/nsb/members/current_members/ochoa.jsp. Accessed 27 Jan. 2021.

"Ellen Ochoa Quotes." Inspiring Quotes, n.d. www.inspiringquotes.us/quotes/uDm4_tPfRyD5V. Accessed 29 June 2021.

"NASA Astronaut Dr. Ellen Ochoa." *NASA,* 29 May. 2018, https://www.nasa.gov/centers/johnson/about/people/orgs/bios/ochoa.html. Accessed 27 Jan. 2021.

Cecilia Payne

Crockett, Christopher. "Cecilia Payne-Gaposchkin Revealed Stars' Composition and Broke Gender Barriers." *Science News,* 1 Mar. 2020, https://www.sciencenews.org/article/cecilia-payne-gaposchkin-revealed-stars-composition-broke-gender-barriers. Accessed 27 Jan. 2021.

Payne-Gaposchkin, Cecilia, and Katherine Haramundanis. *Cecilia Payne-Gaposchkin: an Autobiography and other Recollections.* Cambridge, Cambridge University Press, 1984.

Made in the USA
Monee, IL
05 November 2026

67378008R00029